CLIMATE CHANGE

Andrew J. Milson, Ph.D.
Content Consultant
University of Texas at Arlington

Acknowledgments

Grateful acknowledgment is given to the authors, artists, photographers, museums, publishers, and agents for permission to reprint copyrighted material. Every effort has been made to secure the appropriate permission. If any omissions have been made or if corrections are required, please contact the Publisher.

Instructional Consultant: Christopher Johnson, Evanston, Illinois

Teacher Reviewer: Mary Trichel, Atascocita Middle School, Humble, Texas

Photographic Credits

Front Cover, Inside Front Cover, Title Page
©Ralph Lee Hopkins/National Geographic Stock. **4** (bg) ©Ralph Lee Hopkins/National Geographic Stock. **6** (bg) ©Barry Lewis/In Pictures/Corbis. **7** (tl) Precision Graphics. **8** (bg) Mapping Specialists. **10** (bg) ©Stuart Westmorland/Corbis. **13** (bg) ©David Hiser/Stone/Getty Images. (b) Precision Graphics. **14** (bg) ©Gary Braasch/Corbis. (tl) ©David Evans/National Geographic Stock. **16** (bg) ©Amy Toensing/National Geographic Stock. **18** (bg) ©Ashley Cooper/Corbis. **20** (bg) ©Idealink Photography/Alamy. (tr) ©Christopher Groenhout/Lonely Planet Images/Getty Images. **22** (bg) ©Armin Rose/Alamy. (bl) ©Fiona Stewart/Oceans 8 Productions. **24** (t) ©Fiona Stewart/Oceans 8 Productions. (br) ©Ronald Karpilo/Alamy. **27** (t) ©AP Photo/Jennifer Knight. **28** (tr) ©Brand X Pictures/Jupiterimages. **30** (br) ©David Hiser/Stone/Getty Images. (tr) ©Stuart Westmorland/Corbis. **31** (bg) ©Don Farrall/Photodisc/Getty Images. (bl) ©Gary Braasch/Corbis. (br) ©Amy Toensing/National Geographic Stock. (tr) ©Christopher Groenhout/Lonely Planet Images/Getty Images.

MetaMetrics® and the MetaMetrics logo and tagline are trademarks of MetaMetrics, Inc., and are registered in the United States and abroad. The trademarks and names of other companies and products mentioned herein are the property of their respective owners. Copyright © 2010 MetaMetrics, Inc. All rights reserved.

For permission to use material from this text or product, submit all requests online at www.cengage.com/permissions.

Further permissions questions can be emailed to permissionrequest@cengage.com.

Visit National Geographic Learning online at www.NGSP.com.

Visit our corporate website at www.cengage.com.

Printed in the USA.

RR Donnelley, Menasha, WI

ISBN: 978-07362-97905

14 15 16 17 18 19 20 21 22

10 9 8 7 6 5 4 3

Earth's
CHANGIN
CLIMATE

HOW DOES CLIMATE CHANGE AFFECT LIFE ON EARTH?

Polar bears make their homes far from people. They live in the frozen Arctic and use floating sea ice as platforms to hunt seals. However, the polar bears' icy home is changing. The sea ice on which they depend is shrinking. What's the reason? The answer is **climate change**. Climate change is a long-term change in average weather conditions. The specific type of climate change is called **global warming**. Global warming is a rise in average temperature near Earth's surface. Today's climate change affects more than just polar bears. It affects all life on Earth.

Arctic sea ice is forming later in the year and melting earlier. As a result, polar bears have a shorter hunting season. If sea ice continues to shrink, two-thirds of polar bears could disappear by 2050.

SIGNS OF CLIMATE CHANGE

Over the past 100 years, Earth's average temperature has risen about 1.4° Fahrenheit. This may not seem like much, but this rise is having global effects. It is changing weather patterns and causing animal habitats to disappear. Polar ice masses called glaciers are shrinking.

Although most scientists agree the planet is warming, some people deny it or disagree on *why* it's happening. Earth's climate has changed many times in the past. Is the current climate change part of a natural cycle, as some people claim?

In the United States almost one-third of greenhouse gas emissions come from transportation vehicles, mostly cars.

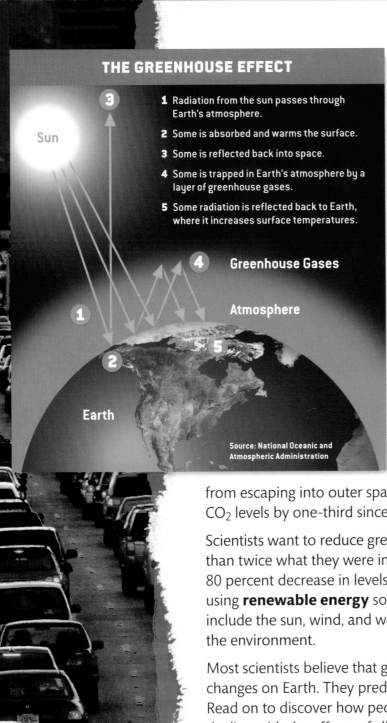

THE GREENHOUSE EFFECT

Sun

1 Radiation from the sun passes through Earth's atmosphere.

2 Some is absorbed and warms the surface.

3 Some is reflected back into space.

4 Some is trapped in Earth's atmosphere by a layer of greenhouse gases.

5 Some radiation is reflected back to Earth, where it increases surface temperatures.

4 Greenhouse Gases

Atmosphere

Earth

Source: National Oceanic and Atmospheric Administration

WHY IS EARTH WARMING?

Most scientists do not think today's climate change is normal. They blame global warming on human activities, especially the burning of **fossil fuels**. Fossil fuels, such as coal and oil, form deep in the earth from decaying plants and animals. They give off carbon dioxide (CO_2) when burned. These fuels power electricity plants, businesses, homes, and vehicles.

Burning fossil fuels puts **greenhouse gases** into the atmosphere. Greenhouse gases are chemical compounds that trap heat near Earth's surface. Similar to the glass roof on a greenhouse, the gases keep heat from escaping into outer space. Human activities have increased CO_2 levels by one-third since the mid-1700s.

Scientists want to reduce greenhouse gas levels to no more than twice what they were in the 1800s. This requires a 50 to 80 percent decrease in levels. To do this, they recommend using **renewable energy** sources. Renewable energy sources include the sun, wind, and water. These sources do not harm the environment.

Most scientists believe that global warming is causing weather changes on Earth. They predict even more changes to come. Read on to discover how people in Costa Rica and Australia are dealing with the effects of climate change.

Explore the Issue

1. **Analyze Causes** How do greenhouse gases contribute to global warming?

2. **Analyze Effects** How is global warming affecting Earth's coldest regions?

WORLD Hot Spots

Climate Challe

NORTH AMERICA
NORTH ATLANTIC OCEAN
NORTH PACIFIC OCEAN
SOUTH AMERICA
SOUTH PACIFIC OCEAN

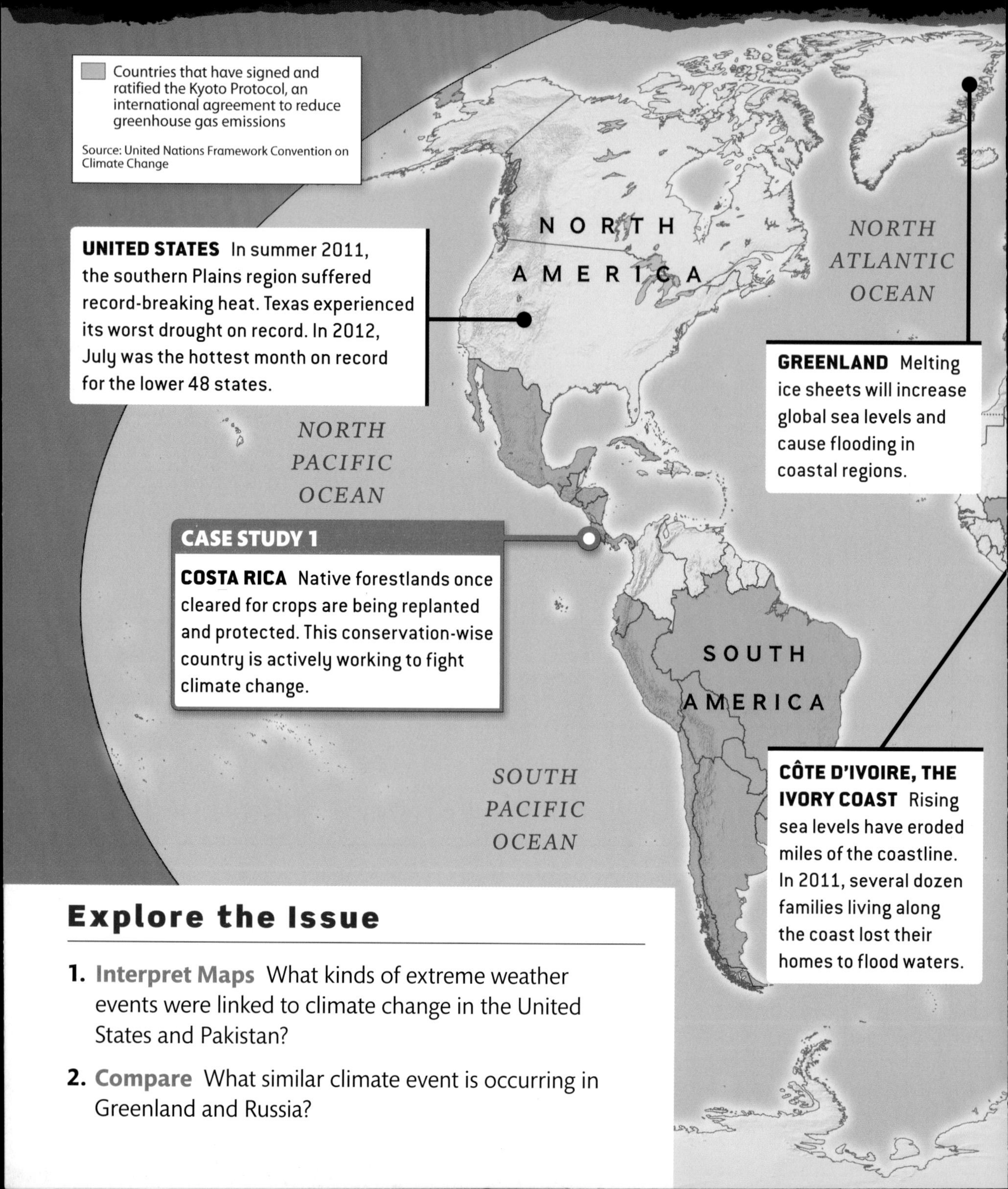

Countries that have signed and ratified the Kyoto Protocol, an international agreement to reduce greenhouse gas emissions

Source: United Nations Framework Convention on Climate Change

UNITED STATES In summer 2011, the southern Plains region suffered record-breaking heat. Texas experienced its worst drought on record. In 2012, July was the hottest month on record for the lower 48 states.

GREENLAND Melting ice sheets will increase global sea levels and cause flooding in coastal regions.

CASE STUDY 1

COSTA RICA Native forestlands once cleared for crops are being replanted and protected. This conservation-wise country is actively working to fight climate change.

CÔTE D'IVOIRE, THE IVORY COAST Rising sea levels have eroded miles of the coastline. In 2011, several dozen families living along the coast lost their homes to flood waters.

Explore the Issue

1. **Interpret Maps** What kinds of extreme weather events were linked to climate change in the United States and Pakistan?
2. **Compare** What similar climate event is occurring in Greenland and Russia?

Study the map. Read the callouts to learn more about the effects of climate change around the world.

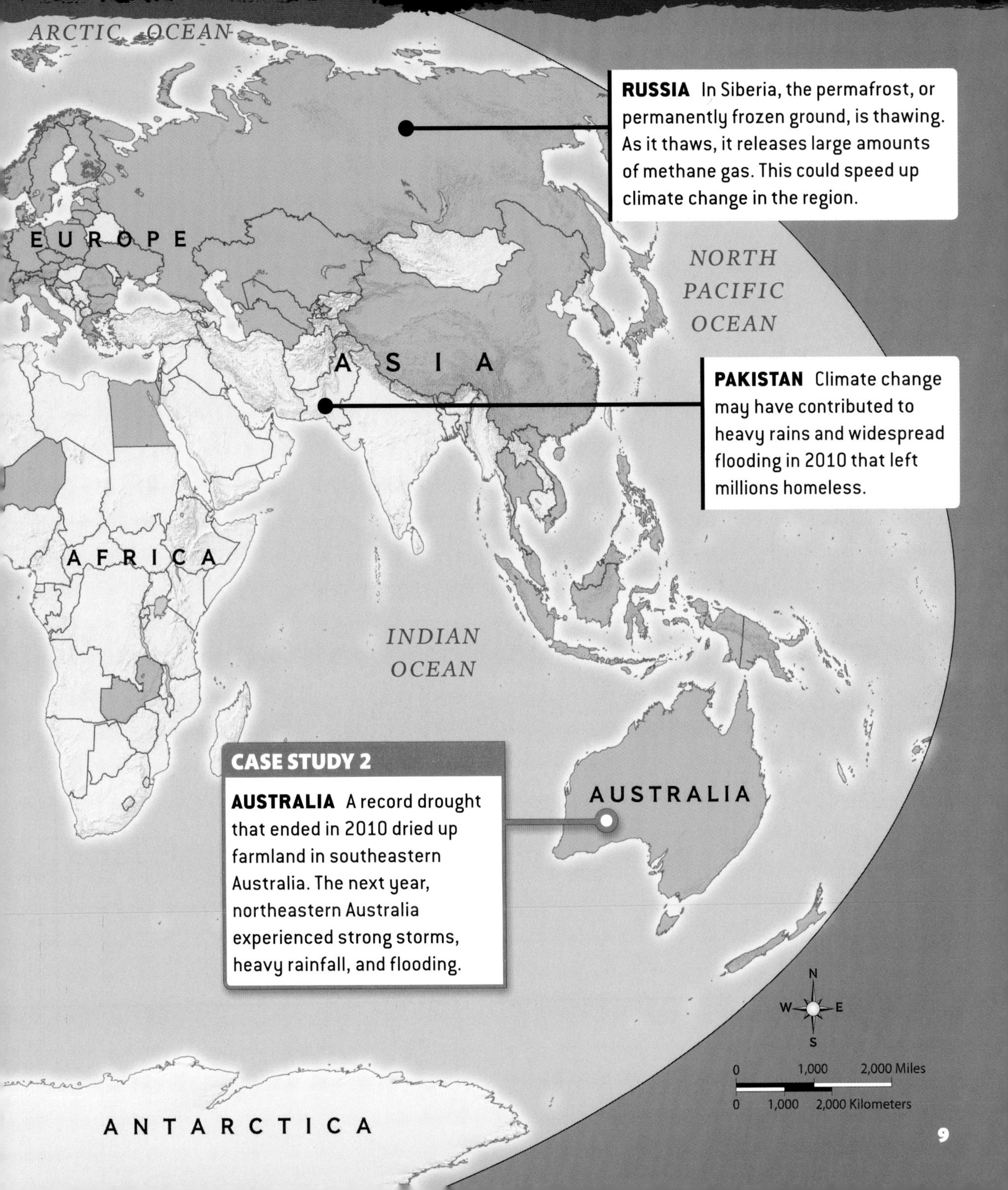

ARCTIC OCEAN

EUROPE

ASIA

AFRICA

AUSTRALIA

ANTARCTICA

NORTH PACIFIC OCEAN

INDIAN OCEAN

RUSSIA In Siberia, the permafrost, or permanently frozen ground, is thawing. As it thaws, it releases large amounts of methane gas. This could speed up climate change in the region.

PAKISTAN Climate change may have contributed to heavy rains and widespread flooding in 2010 that left millions homeless.

CASE STUDY 2

AUSTRALIA A record drought that ended in 2010 dried up farmland in southeastern Australia. The next year, northeastern Australia experienced strong storms, heavy rainfall, and flooding.

N
W   E
S

| 0 | 1,000 | 2,000 Miles |
| 0 | 1,000 | 2,000 Kilometers |

saving Costa Rica's FORESTS

Hikers on a suspension bridge explore the Monteverde Cloud Forest Preserve in Costa Rica. Conservationists work to protect and preserve the country's natural resources.

A NEW DIRECTION

As a young man, Eladio Cruz (il-LAH-dee-oh KROOZ) loved to cut down trees in the rain forests of Costa Rica. He says cutting trees made him happy. "Without a doubt, I was a great destroyer of nature," says Cruz. But today, Cruz is a **conservationist**. He works to preserve forests and wildlife.

Cruz is helping expand the Monteverde (mahn-tiv-VURR-dee) Cloud Forest Reserve, where he once cut trees. **Cloud forests** are a type of rain forest covered by clouds most of the time. Cruz does not think it is easy to restore rain forests. "I think it would take many years to repair the damage I've done," he says.

Cruz's personal story reflects the story of his country. Costa Rica changed from cutting trees to planting them. In the process, the country has become a model for protecting and restoring rain forests. And, as you will learn, forests are one key to slowing the pace of climate change.

LAND OF DIVERSITY

In Costa Rica, coastal land hugs the shores of the Pacific Ocean and Caribbean Sea. Then the land gradually rises to mountains that are almost 10,000 feet above sea level in the center of the small country.

About half of Costa Rica is covered with rain forests. These forests receive at least 100 inches of rain each year. At elevations above 3,000 feet, the rain forests become cloud forests. Water drips from the clouds, providing moisture even when it doesn't rain much.

Costa Rica's geography makes it rich in **biodiversity**, a large number and variety of wildlife and plants. Costa Rica has hundreds of plants and animals that are not found anywhere else on Earth.

To save its unique natural resources, Costa Rica passed a biodiversity law. It protects forested land and pays property owners for conserving the land. The law recognizes that trees help reduce greenhouse gases by absorbing harmful carbon dioxide. In addition, forests provide habitat for wildlife and plants. Cruz and other conservationists hope all people will realize it's not too late to save the rain forests.

FORESTS AND CLIMATE CHANGE

At one time, forests almost completely covered Costa Rica. However, after World War II, people and companies began to cut down the trees for timber and to create farmland. Many other countries, including Brazil and Indonesia, did the same thing. The result was **deforestation**, which is the process of cutting and clearing away trees.

According to most scientists, deforestation is one cause of climate change. To understand why, it is important to understand how trees function. They take in carbon dioxide, a greenhouse gas. During a process called photosynthesis, trees use sunlight to turn carbon dioxide and water into sugar, which trees use for food. They store the carbon in their trunks and leaves. They also release life-sustaining oxygen into the atmosphere.

Through this process, forests absorb more than one-quarter of all the carbon emissions released into the atmosphere. That's an amount equal to the carbon dioxide released by all the cars and trucks in the world. When trees are cut down, carbon dioxide is released into the atmosphere. In addition, there are fewer trees to store carbon dioxide. This, in turn, leaves greater amounts of CO_2 in the atmosphere as a greenhouse gas. For these reasons, according to most scientists, deforestation contributes to climate change.

FORESTS CLEARED FOR FARMLAND

Agriculture was the main reason for deforestation. Using a process called slash-and-burn agriculture, owners of plantations and ranches cleared forests for crops and pasture land. Small farmers also slashed down trees and burned them to clear a few acres for their crops. Loggers cut trees for wood and paper, products important to Costa Rica's economy. In addition, whole forests were cleared to make room for expanding cities.

Like other countries in the region, Costa Rica was destroying its rain forests at an alarming rate. Between 1940 and the early 2000s, Costa Rica's forest cover dropped from 85 percent to 35 percent. Today the National Parks Service protects two-thirds of the remaining forests and teaches students about the importance of preservation.

Slash Forests are cleared to plant crops. Farmers slash, or cut down, trees for farmland.

For centuries, perhaps as far back as Mayan times, farmers in Central America used slash-and-burn agriculture to clear forests.

SLASH-AND-BURN AGRICULTURE

Burn Fallen trees and foliage are burned to clear the land. Ash produced by the fires is used as fertilizer.

Fertilize and Plant Cleared land is fertilized with ash. Crops such as corn and sweet potatoes are planted.

Migrate Groups move on to new locations after soil on cleared lands becomes less productive.

Demand has increased for foods grown organically, or naturally. These coffee beans were produced without the use of pesticides or harmful additives.

Costa Rican foresters plant tree saplings on a barren hillside. The trees will stabilize the soil and prevent erosion.

RESTORING THE RAIN FORESTS

Costa Rica has become a pioneer in **reforestation**, the process of replanting and restoring forests. In one conservation project begun in 1992, land that had been trampled by cattle for 50 years was turned into a lush forest again. People planted seeds collected from native trees that are fast-growing species. Within 5 years, the leafy branches of the trees formed a shady covering over the ground. Soon, other trees began to grow underneath them. Now loggers and local farmers harvest just a few trees at a time so the forest will last.

Through reforestation, the country has regained almost 160 square miles of forest. By absorbing carbon dioxide, the trees help slow the pace of climate change. They reduce erosion and provide a habitat for Costa Rica's extraordinary wildlife. Rain forests also improve the quality of people's drinking water because tree roots are like sponges that clean the water.

Reforestation also boosts the economy. Today many tourists flock to the country's beautiful national parks. **Ecotourism** has become extremely popular. This is travel to natural places to observe wildlife and learn about the environment. In addition, many scientists and students from other countries come to Costa Rica to study tropical forest conservation.

COMBATING CLIMATE CHANGE

Besides preserving forests, Costa Ricans combat climate change in other ways. For example, local students recycle cooking oil waste from restaurants. They use it to make **biofuel**, or fuel that comes from plant or animal waste. The fuel now runs their school buses, replacing diesel gasoline. The country has set a goal of reducing its net carbon emissions to zero by 2021.

The country's constitution states, "Every person has the right to a healthy and ecologically balanced environment." Perhaps a small country will lead the way for the whole planet.

Explore the Issue

1. **Identify Problems and Solutions** How is Costa Rica fighting climate change?

2. **Draw Conclusions** Why should forests be preserved?

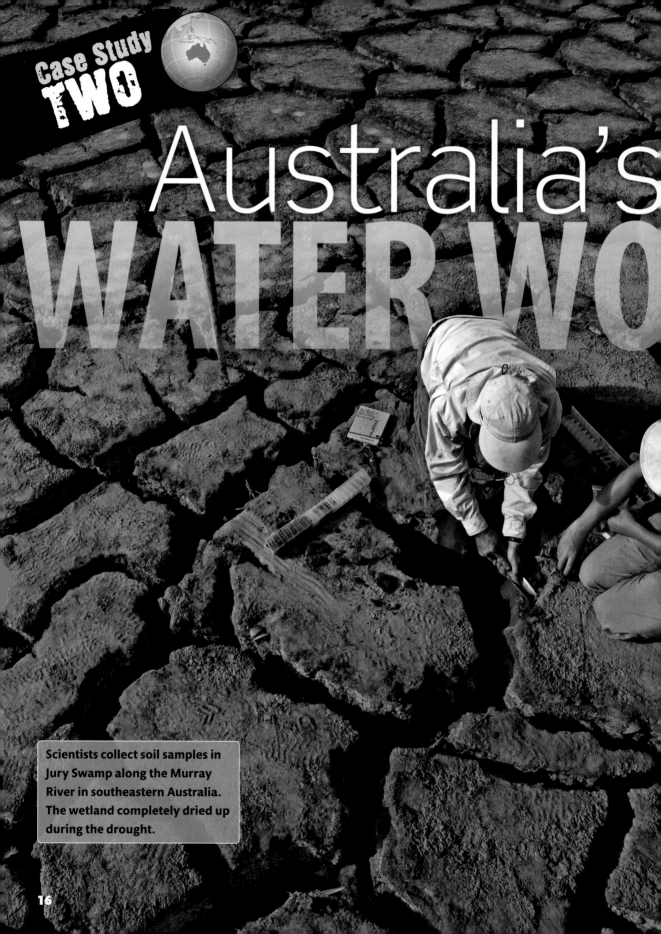

Australia's
WATER WO

Scientists collect soil samples in Jury Swamp along the Murray River in southeastern Australia. The wetland completely dried up during the drought.

DROUGHT WIPES OUT FARMERS

In 2009, farmers who depended on Australia's Murray River were getting desperate for water. They were experiencing the worst drought—an extended period of no rain—in 100 years. In parts of southeastern Australia, the drought had lasted more than 10 years.

Many farmers saw a lifetime of work drying up in front of their eyes. One dairy farmer had only 70 cows left out of a herd of 500. A citrus grower plowed up half of his orchard because he didn't have enough water for his crop. A rice farmer had been growing the water-loving grain since 1962. The farmer had tried to conserve water but finally had to cut back the number of acres he could farm.

Climate change affects people's lives. Australia has been warming—the continent's average temperature has risen by almost 2 degrees since the 1950s. However, Australians are taking steps to deal with the new reality.

WORKING AGAINST NATURE

Getting by with less water is hard on people, animals, and crops. Australia is the world's driest inhabited continent. It has arid and semiarid climate, which means it gets only 10 to 20 inches of rainfall a year. Typically, wet periods alternate with dry periods.

When European settlers arrived in the 1800s, they were lucky to settle during a period of heavier rainfall. The area around the Murray River seemed like a perfect place for farming. Settlers cleared the forests and established dairy farms. They started to grow crops that required a lot of water. But the crops turned out to be ill-suited to the climate. When drier periods came, people used water from the Murray River to irrigate their land. Over the years, too many people tapped into the river. By the 1990s, there was not enough water to go around.

To make the problems worse, the area around the Murray River started getting less rain in 2002. The river's water level dropped so low that salt water from the ocean backed up into the river's mouth and lake. Many freshwater fish and bird species vanished. This event showed how severely a change in the climate can affect the wildlife of a natural area.

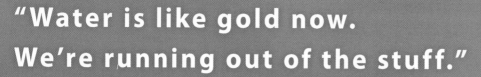

"Water is like gold now. We're running out of the stuff."
—David Harriss, Australian government official

Mining machines near Newcastle—the world's largest coal port—in New South Wales (NSW), Australia, extract coal from an open mine. Australia relies on fossil fuels to produce 75 percent of its electricity.

WEATHER EXTREMES

Climate change has affected different parts of Australia in different ways. For example, in 2011 in Queensland, a state in northeastern Australia, people had the opposite problem from drought—too much rain. Queensland suffered through months of storms and heavy rains. Matthew England of the Climate Change Research Center explained that the waters of Australia were warming and adding moisture to the atmosphere. This can trigger more severe storms.

CARBON DIOXIDE EMISSIONS, 2009

TONS OF CARBON DIOXIDE PER PERSON

The graph shows that countries differ greatly in the amount of carbon dioxide that they produce per person.

Source: International Energy Agency, 2011

Cycles of dry and wet periods have always been part of Australia's climate. However, many scientists believe that as Earth's atmosphere grows warmer, the swings from wet weather to dry weather will be more extreme. Speaking in *National Geographic* magazine in April 2009, Isaac Held, a scientist for the U.S. National Oceanic and Atmospheric Administration, said, "Wet areas are going to get wetter and dry areas drier."

Some scientists do not think the changes in Australia's climate will be permanent. They believe the events might be variations in a natural cycle. Other scientists strongly disagree, saying eastern and southern Australia will continue to see less rainfall. They also project temperatures might climb 0.6 to 1.5 degrees Celsius by 2030. The swings between drought and heavy rains could grow even more severe.

If this happens, Australia will find it even harder to meet the water needs on farms and in cities in the future. During the recent drought, many cities had to restrict people's use of water. People hauled buckets of used shower water to their gardens. They installed artificial lawns, which do not need to be watered. At the end of the latest drought, David Harriss of the NSW Office of Water in Sydney said, "Water is like gold now. We're running out of the stuff." Without question, the droughts are changing how Australia uses its water.

Large solar collectors track the sun and convert its heat to mechanical energy.

A driver tests a solar-powered test car. Vehicles powered by a renewable energy source do not harm the environment.

A NEW ENERGY FUTURE

Most experts blame Australia's dependence on coal for rising temperatures and changing rainfall patterns. For the size of its population, Australia releases a large amount of CO_2 into the atmosphere. Most of the country's carbon emissions come from burning coal, which is a major source of carbon dioxide.

In 2011, Australia introduced the Clean Energy Future plan. Its goal is to protect the environment from damage due to climate change. The government wants to make its industry more **sustainable**, or able to continue into the future with the least amount of harm to the environment.

A main feature of Australia's energy plan is a tax penalty paid to the government for carbon emissions. The country's 500 largest polluters will have to pay a tax equal to their carbon emissions. Australia thinks this tax will lead businesses to reduce pollution and use less energy. Money from the tax will be used to help customers switch to renewable energy sources, such as solar and wind power.

Australia has set high goals for its energy plan. It wants to cut carbon pollution by 23 percent in 2020. That would be equal to taking 45 *million* cars off the road. The country plans to close older power plants and produce much more electricity from renewable energy sources.

Because of its sunny climate, Australia started a Solar Cities Program, which helps businesses and homes use energy wisely. The goal of the program is to reduce the use of fossil fuels and increase the use of solar energy. Adelaide, one of seven cities in the program, introduced the first solar electric bus in the world. Western Australia's largest solar site includes eight buildings at the Perth Zoo. Thousands of schools around the country have changed to solar power. Clearly, Australia is taking important steps to fight climate change.

Explore the Issue

1. **Summarize** How is climate change affecting Australian farms?

2. **Identify Problems and Solutions** How is Australia fighting climate change?

Exploring Antarctic Ocean

Under sail to King George Island, Antarctica, explorer Jon Bowermaster crosses Drake Passage, one of the windiest places on Earth.

TAKING EARTH'S PULSE

Explorer Jon Bowermaster has seen the effects of climate change all over the world. He was awarded grants by the National Geographic Expeditions Council to develop the Oceans 8 project. Since 1999, he has explored the oceans of the world and recorded his work in documentary films and in articles in magazines such as *National Geographic*.

Bowermaster has led expeditions to all seven continents to study the oceans. The oceans actually form one vast sea with different names in different places. They cover more than 70 percent of Earth's surface and absorb more than 80 percent of the heat trapped by greenhouse gases. Almost half the world's population depends on the oceans for food, transportation, and energy. Climate change in the oceans affects billions of people and marine life everywhere.

CHECKING EARTH'S HEALTH

Antarctica's extreme climate closes the continent to exploration most of the year. Giant, slow-moving mountains of snow and ice called **glaciers** and massive layers of thick glacial ice called **ice sheets** cover Antarctica year round. During its long winters, new ice forms and doubles the surface of the continent. This natural expansion provides a suitable habitat for penguins, seals, and seabirds but not for people. Explorers must wait for the short summer months to explore Antarctica's ocean.

In January 2008, Bowermaster and his team set sail for West Antarctica. Just like the penguins in the region the explorers quickly learned to steer around icebergs. When floating chunks of ice collide, even large ships with reinforced hulls can become trapped in ice jams. So, the team launched kayaks from the transport ship and paddled toward the peninsula. Their plan required nerves of steel. Floating ice can easily capsize lighter craft.

As the team explored the waters around the peninsula, they found evidence of climate change. "If you believe as I do that Earth is a living thing, then Antarctica is its pulse," Bowermaster says. Climate change that occurs on the most remote continent affects all life on Earth.

Towering icebergs rise above the ocean waters near the Antarctic Peninsula.

In light, maneuverable kayaks, Oceans 8 explorers paddle around floating icebergs off the Fish Islands, Antarctica.

STUDYING THE WATERS

Bowermaster believes that the ocean around Antarctica shows the effects of climate change. The waters contain only about 5 percent of the world's seawater, but they are home to 20 percent of its sea life. Rising temperatures are changing the patterns of the wildlife. Bowermaster noted a decline of Antarctic wildlife, including penguins and krill, the tiny shrimplike crustaceans that feed larger sea animals. He also noted an increase in water plants.

During their exploration, Bowermaster and his team saw a leopard seal asleep on a floating piece of ice and some rare white penguins. However, human activity may soon affect the region. He saw over a dozen tourist ships in the area. Bowermaster says that a decade ago, "if you saw one private sailboat here it was amazing." Local scientists will monitor the impact of this increased tourism.

These photos of Muir Glacier in Alaska were taken 100 years apart. They show the effects of global warming in another part of the world.

"If you believe as I do that Earth is a living thing, then Antarctica is its pulse." —Jon Bowermaster

KEEPING WATCH

Bowermaster was also surprised by the amount and thickness of ice in Antarctica's ocean. During the previous winter, Antarctica had experienced colder than usual temperatures. However, Bowermaster points out that thick ice is due to seasonal weather. It should not be taken as a sign to ignore the effects of climate change the Oceans 8 team observed.

On his trip, Bowermaster learned that Earth's coldest continent needs to be watched closely. As he says, "If Antarctica is the beating heart of the planet, as I believe it is, it deserves to be cared for with all of our best intentions."

Explore the Issue

1. **Explain** According to Bowermaster, why is it important to keep a close watch on Antarctica's climate?

2. **Draw Conclusions** Why do the effects of climate change in Antarctica matter to the rest of the world?

Plant a Tree
—and report your results

You don't have to explore the oceans to fight climate change. You might start by cutting down on your use of fossil fuel energy. You've also learned that trees help reduce the amount of carbon dioxide in the atmosphere. By planting trees in your community, you can make a difference.

IDENTIFY

- Find out about tree-planting projects in your school or community.

- Invite experts to explain the best kinds of native trees for your climate. Also learn when and how to plant saplings.

- Find a donor source for trees. Set a date for their delivery.

ORGANIZE

- Advertise your project through school or local media.

- Have volunteers sign up to plant the trees and gather supplies, such as gloves, shovels, water, and mulch, for planting day.

- Schedule volunteers to water the newly planted trees weekly.

Students work together to plant a tree in their community on Arbor Day.

DOCUMENT

- Take before-and-after photos of the site(s) where you plant the trees and photograph the trees during each season.

- Label your trees with signs or make a map of the area that shows the kinds of trees you have planted.

- Keep a log about when and how you care for the trees, and record their growth each month.

SHARE

- Use your photos and recordings to create a multimedia presentation about your tree-planting project and share it with your class.

- Create a blog about your project from planning to planting and follow-up. Include information about trees and climate change.

- Encourage others to plant trees by creating a visual display of your project for your local library or community center.

Research & WRITE
Narrative

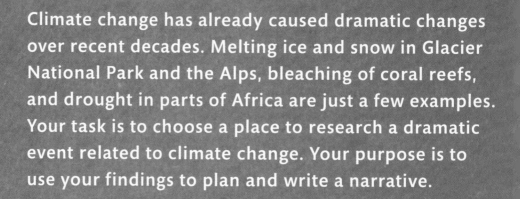

Write a Narrative

Climate change has already caused dramatic changes over recent decades. Melting ice and snow in Glacier National Park and the Alps, bleaching of coral reefs, and drought in parts of Africa are just a few examples. Your task is to choose a place to research a dramatic event related to climate change. Your purpose is to use your findings to plan and write a narrative.

RESEARCH

Use the Internet, books, and articles to find the following information:

- What was the climate like in the early 1900s?
- How has the climate changed over time?
- What dramatic natural events occurred, and how are they related to climate change?

As you research, be sure to take notes and keep track of your sources.

DRAFT

Review your notes about the location you chose. Then select one climate event for your story.

- Begin the first paragraph by describing the event's setting with vivid details that will engage the reader.
- Narrate the story through the eyes of a person or an animal. Establish and maintain a consistent point of view, and provide the context for your story.
- The middle paragraphs should present the sequence of events that led to the changes. Use precise words and sensory language to describe them. Use helpful transitions to convey sequence.
- In the final paragraph, wrap up the story in a way that makes the reader want to learn more.

REVISE & EDIT

Read your first draft to make sure that it clearly tells the story of how climate change affected one place in time.

- Does the beginning get the attention of your audience and introduce the topic and setting?
- Does the body of the story present a logical sequence of events and include descriptive details?
- Does the ending flow from the story in a reflective, thought-provoking way?

Revise the narrative to make sure it is clear and complete. Then check your paper for errors in grammar, spelling, and punctuation.

PUBLISH & PRESENT

Now you are ready to publish and present your narrative. Print out your narrative, or write a clean copy by hand. Consider adding photos or graphics to enhance the text. Publish your narrative in a class magazine or Web site.

Visual GLOSSARY

biodiversity *n.*, a large number and variety of wildlife and plants

biofuel *n.*, fuel from plant or animal waste

climate change *n.*, a long-term change in average weather conditions on Earth

cloud forest *n.*, a type of rain forest covered by clouds most of the time

conservationist *n.*, someone who works to preserve natural environments and wildlife

deforestation *n.*, the process of cutting and clearing trees for fuel or farmland

ecotourism *n.*, travel to observe wildlife and learn about the environment

fossil fuel *n.*, a carbon-based energy source that forms in the earth

glacier *n.*, mountains of snow and ice that cover the land in Earth's coldest regions

global warming *n.*, a rise in the average temperature near Earth's surface

greenhouse gas *n.*, a chemical compound released into the atmosphere where it traps heat near Earth's surface

ice sheet *n.*, a massive layer of glacial land ice in Earth's polar regions

reforestation *n.*, the process of replanting trees and restoring forests

renewable energy *n.*, a natural power source that does not harm the environment

sustainable *adj.*, able to continue conservation practices into the future

biodiversity

deforestation

renewable energy

reforestation

climate change

INDEX

SKILLS